1

"If it could be demonstrated that any complex organ existed which could not possibly have been formed by numerous, successive, slight modifications, my theory would absolutely break down." – Charles Darwin. (*The Origin of Species*, 1859).

THE DEMISE

OF

DARWINISM

OR

THE RISE AND FALL OF

THE THEORY OF EVOLUTION

By

H. Clay Gorton

Edited and with a preface by Stephen R. Gorton

Cover Design by Connie Gorton

ISBN 978-0-9852470-1-0

Editor's Note: With the exception of a few minor corrections
in punctuation and spelling, this manuscript is being
published in its original version as written by the author.

www.sgorton.com

To Podge

PREFACE

"In the long search for truth, we have often reached wrong conclusions both in science and religion....Life demands that we keep learning. Otherwise we become shunted off into dark pockets of error." (Franklin S. Harris, *Science and Your Faith in God,* Bookcraft, 1958, p 93).

Scientific method entails supposition and conjecture. Without exception, these always encompass the probability of error. History chronicles countless scientific concepts created on incorrect and erroneous suppositions. Most of us can, without difficulty, bring to mind so called "scientific facts" which have now been demonstrated to be incorrect, inaccurate or flawed.

Some of the obvious assumptions include:

The world is flat. Earth is the center of the universe. Chemical elements cannot be divided into smaller particles. Newton's laws explain all forms of motion. Outer space is filled with ether.

Less dramatic assumptions in the field of science have led to even larger blunders than those mentioned above.

"Science is *not* a body of indisputable and immutable truth," wrote Anthony Standen in his book *Science is a Sacred Cow.* (E.P. Dutton & Son, Inc., New York, 1950). "It is a body of well-supported probable opinion only, and its ideas may be exploded at any time."

When our opinions are based on a false assumption, we can defend all kinds of mendacities as true and accurate. Incorrect suppositions can seem to progress plausibly and even appear accurate. Time has shown that hypotheses fabricated on false suppositions will eventually crumble but an assertion originating from a true and accurate assumption will persevere.

Suppositions take a pronounced position both in science and in religion. Certain religious principles can be founded on false beliefs and yet appear truthful and reasonable. It is as easy and, indeed, just as common in religion to espouse a false notion as it is in science, conceivably even more so. Science, in general, is more proficient than religion at examining and extinguishing fallacious premises. In science, when a theory is observed to be incorrect, it is usually rescinded. Religion is somewhat less inclined to consider false concepts and false traditions among its beliefs and teachings.

As stated above by Franklin S. Harris, we should never stop learning. This is as equally true in religion as it is in science. The views most people have of religion were acquired in childhood. Much of the world's opinion regarding religion is founded on kindergarten concepts. When we fall behind in our religious learning, religion appears irrational. Many have abandoned religion simply because they failed to develop their spiritual knowledge along with their temporal knowledge. Only immature religion is unintelligent.

Some individuals feel that, in this Age of Enlightenment, established religion cannot endure the onslaught of scientific discovery. In his book, *Issues in Evolution,* (University of Chicago Press, 1960, p. 252). Sir Julian Huxley wrote, "In the evolutionary pattern of thought there is no longer either need or room for the supernatural. The earth was not created; it evolved. So did all the animals and plants that inhabit it, including our human selves, mind and soul as well as brain and body. So did religion."

In contrast to this viewpoint, others maintain that there has been an abundance of evidence to support the existence of God which parallels and even surpasses the surge of scientific advances.

Any rational, thinking person would be obliged to investigate the evidence for the existence of God. Unfortunately, average intellects are disinclined to consider issues that present a threat to their comfortable and convenient existence.

"Come now," counsels the prophet Isaiah, "and let us reason together." (Isaiah 1:18). On one side of the argument there are people who allege to be atheists or agnostics; on the other are those who affirm that they "know" God lives. Is either being misled or deceived?

Seven significant statements to consider when examining *"The Demise of Darwinism"* are the following:

1. When they discard their false ideologies, no conflict will exist between exact science and true religion. The disagreement is not between science and religion but between two differing philosophies, one of which hides behind a pretense of scientific propriety.

2. Both faith and reason have proven their importance in human experiences. Faith plays a significant role in science as well as in religion, and reason is needed just as much in religion as it is in science.

3. Science and logic can never conclusively prove the fundamental principles of faith, which include the reality of a living God, the divine nature of our Savior Jesus Christ, and our own immortal existence.

4. Thanks to science we live in an astonishingly superior world than our antecedents. Science and religion can both endure as abundant sources of life.

5. Although science and religion principally function in distinct areas of life, they are not mutually exclusive but go hand-in-hand toward a common goal—the search for truth.

6. On-going revelation permits religion to continually assist and benefit contemporary humanity. It supplies the answers and assurances we seek.

7. Obedience to eternal, unchangeable law, whether scientific or spiritual, leads to true freedom and happiness.

There is plenty of "evidence of things not seen" (Hebrews 11:1) and there is plenty of hope for a better world.

Stephen R. Gorton -- 2012

FORWARD

In today's world technology is king. The marvelous advances in technology during the last century are astounding. More advances have been made in the last one hundred years that affect the comfort and convenience of humankind than in all of humanity's history. Things unimaginable one hundred years ago are children's toys today. This has produced in the minds of today's generation, and especially the youth, the concept that science is infallible, and it is an accurate and true description of the world around us.

Comparing science to religion, it is the common argument in today's world that if historical accounts in the Bible are at odds with the pronouncements of science, the biblical accounts are relegated to allegory or mythology. Further, the validity of religious doctrinal topics is also judged on the basis of their conformity with current scientific theories. Such topics include the concept of time and the concept of light, including the nature and speed of light. Perhaps nowhere is the controversy between science and religion more acute than in the fields of archeology, anthropology and paleontology, since they deal with the conflicting biblical timelines of the creation of the earth, the sequence of life forms on the earth, the advent of man and man's expulsion from the Garden of Eden. At the very core of these controversies is the scientific concept of the spontaneous generation of life from primordial ooze, and the religious concept of the creation of life by a Supreme Being.

Further, the theories of science are so deeply ingrained in the minds of academia, that any evidence that does not concur with current theory is simply discarded. *It requires a true revolution in scientific thinking to discard current theories for a new world view.* Those revolutions have repeatedly occurred, and are occurring today.

The purpose of this little book is to point out some of the glaring inconsistencies in the theory of uniformitarianism, of which Charles Darwin's theory of the survival of the fittest and the gradual evolutionary changes over vast periods of time are central themes. Another purpose is to identify some of the outstanding evidences that support the theory of catastrophism, which claims that catastrophic events have been responsible for the earth's present configuration, consistent with biblical accounts. Our objective in publishing this book is to help today's youth, and other decisive thinkers, develop a critical eye regarding the theories of the human race, and to see them for what they are -- merely theories and not absolute truth. Also, it is our hope that the material presented herein will provide a basis for the increased understanding of, and faith in, the Holy Scriptures and in the pronouncements of the prophets of God, both ancient and modern.

INTRODUCTION

Tempest in a Teapot

The controversy between creationism and uniformitarianism has continued to boil since the time of Sir Isaac Newton. One interesting bubble broke the surface in the State of Tennessee in 1925 in which a school teacher, John T. Hoopes, was enjoined by law from teaching the theory of evolution in the public school. He was fined $100 for so doing. The trial, with William Jennings Brian and Charles Darrow as opposing attorneys, attracted national attention and came to be known as the "Monkey Trial". The ruling in this trial forbade the teaching, in any state-funded educational establishment in Tennessee, of "any theory that denies the story of the Divine Creation of man as taught in the Bible, and to teach instead that man has descended from a lower order of animals." This is often interpreted as meaning that the law forbade the teaching of any aspect of the theory of evolution.

The pendulum has now swung to the opposite extreme. In 2005, eighty years after the "Monkey Trial", Senator Chris Buttars (R-West Jordan, Utah) attempted to introduce legislation which would permit the teaching of "Divine Design" in public schools in the State of Utah as an alternative to the theory of evolution. Because of the vociferous objections to teaching Divine Design in public schools voiced by the American Civil Liberties Union of Utah, which erroneously interprets the provision in the U.S. Constitution prohibiting the government from imposing a State religion on the citizens as precluding any governmental institution from supporting any organization or activity that invokes or supports any religious concept, and Utah Academia, which promotes evolution as an established law rather than as a *theory* of science, Senator Buttars has withdrawn his proposed legislation.

Furthermore, in October 2004 in Harrisburg, Pennsylvania, in one of the biggest courtroom clashes between faith and evolution since the 1925 Monkey Trial, a federal judge barred a Pennsylvania public school district from teaching "intelligent design" in biology class, saying the concept is creationism in disguise. U.S. District Judge John E. Jones delivered a stinging attack on the Dover Area School Board, saying its first-in-the-nation decision to insert intelligent design into the science curriculum violates the constitutional separation of church and state.

The Combatants

The combatants in this controversy over the freedom to teach Divine Design in public schools are unevenly divided opponents. The evolutionists, comprised of the academic faculties and other far left organizations and individuals, speak with essentially one voice to keep the concept of God as a creator out of public school curricula. Their weapons include the discovery of fossilized remains of both plants and animals in buried strata all over the world, and their dating techniques rely on the decay rates of radioactive carbon-14 and potassium-40.

The proponents of the concept of God as a creator, on the other hand, speak with many conflicting voices. In general, the major religious groups of Protestants, Catholics, Jews and Mormons are divided among themselves in their beliefs concerning Divine Design. Their weapons are restricted to their interpretations of the Old Testament. The Mormons, however, include in their arsenal, in addition to the Old Testament, other modern scriptures. The Mormons generally speak with one voice, but among the other major religious groups there are numerous diverse interpretations of the meaning of the scriptures that relate to the creation process.

Skirmishes

In the battle over teaching Divine Design in the public schools, the proponents of Divine Design do not have a chance of succeeding. They are as a number of undisciplined, independent combatants on the home turf of a major disciplined army, with vast reserves of intellectual weapons and backed by a powerful government that has enacted laws legitimizing their agenda.

Fighting Fire with Fire

The only way that the evolutionists can ever be overcome in their well-established beliefs is to fight fire with fire. There is the need for re-introducing into the academic environment an intellectual philosophy that would use the same tools and rules as the evolutionists, but which would champion the concepts advocated by the proponents of Divine Design. Such a discipline is embodied in the concepts of catastrophism, a geological doctrine that changes in the earth's crust have been brought about by sudden cataclysmic events rather than by imperceptibly slow uniform changes over time. Rather than using the scriptures to combat the concept of evolution, the proponents of catastrophism would invoke data from prehistoric volcanoes, extensive lava flows, seashells and other marine fossils found in strata at high elevations, coal veins with carbonized trunks of massive trees in wild disarray, extensive shore-line mesas at various elevations worldwide and extensive sedimentary strata in highly modified orientations.

This introduction depicts an episode in the larger juxtaposition between establishmentarianism and innovation in the scientific community since the time of the Greek culture. There is the common belief in academic circles that scientific observations of the external world, when formalized by a logical description and reduced to the language of mathematics, represent a description and explanation of fundamental, universal truth, which exists aloof and independent of the observer. Each succeeding age has had its unique "truths" of the surrounding environment; and the truths of each age have been replaced by those of each succeeding age, as observation and measurement techniques and equipment have been developed and evolved. The theory of evolution is the academic truth of our day. It is held with the same tenacity as have been all the defunct truths of ages past. But in spite of the dogmatism of its adherents it will find its way into the dustbin of history, as have all the erroneous theories of science that have preceded it. In the treatise that follows, we will present the compelling evidences that signal the demise of the theory of evolution.

FUNDAMENTAL CONCEPTS

To begin with it is necessary to clarify a couple of key concepts -- the fundamental postulates of science, and the fundamental postulates of revealed religion. By the term science we refer specifically to the natural sciences including such fields as physics, chemistry, astronomy, geology, paleontology and mathematics; and to the social sciences, including such fields as anthropology and archeology. By revealed religion we refer to all the concepts revealed by both ancient and modern prophets of God to man.

Each of the disciplines of science, without exception, is based on logical developments from *a priori* postulates. *A priori* postulates are the basic assumptions on which each discipline is based, and are stated premises that are accepted as being true without being subject to proof. Other than the *a priori* postulates, the acceptance of a concept as a true fact in science must meet the criteria of being observable, repeatable, and communicable. This procedure invokes the concept of doubt -- nothing is accepted as being a fact until it can be established by the criteria of observation, repetition and communication.

In the discipline of religion, the truthfulness of a concept is established by first assuming the concept to be true, then putting the concept into practice by which its truthfulness is verified. This procedure invokes the concept of faith -- the assumption of truth before the fact.

The Astounding Advancements of Science

During the last century science has made astounding advances in almost all fields. Consider trying to explain to a person who lived a hundred years ago the technology of today. For the first time in history the scientific concepts of today are so far advanced from those of a hundred years ago as to be completely incomprehensible to someone of that time period. These marvelous advances have modified and control every aspect of life in today's world. Among the things that we accept as commonplace today that could not have been understood or imagined by those of the nineteenth century are our communication systems, our information transmittal and storage systems and our power systems, to name a few.

All of these amazing new concepts have been developed by research in the technological industries and in the universities. The universities are the principal promoters, and together with professional societies, are the principal guardians of scientific knowledge. It is important here to note that scientific knowledge, without any exception, is based on *a priori* premises, or *axioms* -- given postulates that are not subject to proof, but represent the basis of the logic that is used in the development of any and all scientific concepts. Formal systems of logic are used to develop concepts using the given *axioms* and theoretical or experimental concepts or data.

THEORIES AND LAWS OF SCIENCE

The concepts developed by scientific procedure are accepted as *theories*. When a given theory has been tested to the point of gaining general acceptance in the scientific community, it is given the title of a *law* of science. The laws of science are accepted a universally true, and are generally not subjected to further examination for verification. Yet these scientific laws are nevertheless still based on *a priori* premises that are accepted as true without experimentation or verification.

To consider just one or two examples of the application of *a priori* premises in the development of scientific knowledge, let us consider first the very exact science of geometry. Euclid (330-270 B.C.) developed the concepts of plane geometry that included three *a priori* postulates; 1) a straight line is the shortest distance between two points, 2) two parallel lines never cross and 3) two non-parallel lines cross at one and only one point.

These three concepts are accepted as true without proof, and form the basis of Euclidian geometry. The concepts appear to be so logical that it may be assumed that they need no proof. However, there are other practical geometries that use other, conflicting *a priori* premises.

Consider Riemannian geometry. George Friedrich Riemann (1826-1866) developed a geometry based on different, and conflicting, *a priori* postulates to those of Euclid. Among Riemann's conflicting assumptions were: 1) the shortest distance between two points is a curve, and 2) there are no lines parallel to a given line. e.g., two seemingly parallel lines cross at infinity.

That Riemannian geometry is a perfectly valid system of logic is demonstrated by the fact that Albert Einstein used Riemannian rather than Euclidian geometry in the development of his theory of relativity. Thus we may ask, is the shortest distance between two points a straight line or a curve? Of course, although conflicting with each other, both concepts are true in the context in which they are defined. However, neither can be said to be absolutely or universally true because they are in conflict with one another. Neither concept is the expression of the observation of reality, but rather each is an invented construct.

Let us now consider the *a priori* concepts in Newtonian mechanics. Newton's basic postulates involve mathematical definitions of mass, time and length. These three concepts appear to be directly related to the senses and to be descriptions of the real world. However, when examined critically, they are all imponderable mathematical concepts.

Mass was given by Newton as the proportionality constant between force and acceleration, and has the appearance of being an intrinsic and invariant property of all matter. Einstein, on the other hand, developed mass as a variable property, whose value depends on the velocity of the object with respect to the observer.

In his definition of time, Newton stated, "We conceive true or absolute time, to flow uniformly in an unchangeable course, which alone serves to measure with exactness the changes of all other things." (Colin Maclaurin, *An Account of Sir Issac Newton's Philosophical Discoveries,* London, 1748, pp 100-101). Newton's concept of an 'absolute' time, which assumes a standard clock somewhere in the universe is not confirmable and therefore loses its justification. Dr. Georg Joos, Professor of Experimental Physics at the Technische Hochschule, Munich, commented on the subject,

> "Nevertheless, the fact that Newton was able to formulate a thoroughly adequate foundation for the mechanics of macroscopic bodies by

introducing these two concepts, (fixed mass and absolute time) *whose difficulties could scarcely have escaped his notice*, is to be regarded as an outstanding stroke of genius." (Georg Joss, *Theoretical Physics,* Hafner Publ. Co., New York, p 227. Italics ours).

The concept of length, or the measurement of distance, seems straightforward. The accepted procedure is to obtain a stick of standard length and lay it repeatedly end-to-end from point A to point B. The number of multiples of stick length between A and B is said to be the intervening distance. The first problem comes when we try to measure the distance from A to B across a river. Since the stick cannot be held in a fixed position on the water, an alternate method must be used. The standard stick is used to measure the distance between points A and C along the shore, and then by triangulation the distance between A and B on the other side of the river is calculated. The calculations, however, utilize the principles of Euclidian geometry, which are based on improvable *a priori* assumptions, and therefore can only be stated to be true in terms consistent with those initial assumptions.

An additional problem arises when we attempt to measure astronomical distances. One of the accepted techniques for measuring the distance between the earth and stars is to infer the Doppler effect of increase in wave length of light emitted from the star as proportional to the speed of recession of the star. This inference is based on the assumption of a uniformly expanding universe such that all celestial bodies are moving away from the earth at a rate proportional to their distance from the earth, in the same manner that ink dots on an inflating balloon all move away from each other at a rate proportional to their distance from one another. At the same time, the concept of an expanding universe comes from the observed Doppler Effect in the wave length of the red spectral lines of light emitted by hydrogen nuclei in the stars. Thus the measurement system is based on the theory of the system being measured!

THE SOCIOLOGY OF SCIENCE

It may be surprising to some to hear that for a scientific theory to be accepted it must first be sociologically acceptable. Each age has had its truths that have fit with the total intellectual and religious climate of that age. The earliest days of what could be considered in a broad sense the scientific method were the period of time when Greece was in power. Up until the Roman conquest Greece was the seat of intellectual learning. In fact, the Greeks of that time developed many of the concepts that have remained until today as fundamental concepts of scientific methodology – witness Euclidian geometry that has already been mentioned. Pythagorus and Archimedes were two other great Greek mathematicians whose concepts remain as fundamental parts of modern-day mathematics. Nevertheless, much of what was true at the time of the flowering of Greek culture is now understood to be completely without foundation.

Among other theories held by the Greek to be true that have found their way into the overloaded dustbins of history is the theory first proposed by Anaximander in 546 BC that the fundamental particles of the universe consisted of air, earth, fire and water. This concept is manifest in the Book of Abraham 2:7:

"For I am the Lord thy God; I dwell in heaven; the earth is my footstool; I stretch my hand over the *sea*, and it obeys my voice; I cause the *wind* and the *fire* to be my chariot; I say to the *mountains*--Depart hence--and behold, they are taken away by a whirlwind, in an instant, suddenly."

This demonstrates that God speaks to men according to their own understanding. (See Doctrine & Covenants 1:24).

Aristotle, born in 384 BC, held that men had more teeth than women. This was a very logical assumption, since men's heads are generally larger than are women's heads. The Greeks would never attempt an experimental verification of this theory because in their culture such menial tasks were not performed by the elite --that was what slaves were for. In addition, in their culture, logic was all that was needed to establish a truth.

The Tenacity of Scientific Belief

It has been the experience of each succeeding era of scientific thought that the free thinkers, those who dared to challenge the scientific establishment, were ostracized by the scientific community and often by society in general, and received their due punishment -- from house arrest to the termination of their existence. Nor were the ancient Greeks free from such establishmentarianism. The great philosopher, Socrates, at age 70 in 399 BC, was condemned to death on trumped up charges of corrupting the youth and interfering with the religion of the city. Sound familiar?

The Geocentric Universe

The Greeks held to the concept of a geocentric universe -- the sun rotated around the earth and the stars were objects on a sphere that also rotated around the earth. However, the planets (from the Greek term for wanderers) were not fixed in position with the other stars. Their motions, however, could be accounted for by assuming that their orbits included epicycles within their circular orbits. By assigning radii and periods of rotation to the assumed orbits their motions could be well accounted for. This truth of that age held sway for some 1,700 years.

The Heliocentric Universe

Nikolai Copernicus (1473-1543) proposed a heliocentric universe that had first been proposed by Aristarchus in Greece in 200 BC, but that did not gain prominence because of the popularity of an earth-centered system. The heliocentric theory met with much resistance by the Catholic Church and others who felt more comfortable, along with the early Greeks, with an earth-centered universe. Galileo (1564-1642) was condemned as a heretic by the Catholic Church for publishing a work, *Dialogue Concerning* the Two Chief World Systems, in which he championed the heliocentric concept of the universe. He was imprisoned, but later released to house arrest because of his very poor health until his death.

The Expanding Universe

Now the scientific community has struggled on to what is popularly known as the Big Bang theory. Because of the Doppler shift in the hydrogen spectral lines from the stars, a constantly expanding universe is deduced. By mathematically reversing the acceleration of the stars away from each other scientists have been able to deduce that the universe began as an infinitesimally small point of infinitely high density and temperature between 10 and 20 billion years ago. (Observations of white dwarfs in the Milky Way Galaxy with the Hubble space telescope by Jet Propulsion Laboratories, Pasadena, California, in 2002 have refined the estimate of the age of the universe to between 13 and 14 billion years). This theory represents the truth of our modern age about the nature of the universe. However, there is one question in that regard that scientists will not entertain -- what was the universe like, say, one millionth of a second before the big bang explosion?

The Velocity of Light

That brings us to the measurement of the velocity of light and to a consideration of the nature of light. Light was thought of by Sir Isaac Newton (1643-1727), the founder of the physics of macroscopic bodies, to be composed of tiny particles. Were this the case, as light would travel through dense media it would necessarily be slowed down. Newton could mathematically account for the apparent bending of a stick partially immersed in water at the air/water interface by the slowing down of light as it passed through the water.

When light was passed through a very small hole or a narrow slit, however, its direction was observed to be modified in a regular and ordered pattern. Such a phenomenon could not be explained by the particulate theory of light. However, if light were considered to be a wave in some medium that propagated the light energy, then the observed variations from a linear transit through a small hole or slit could be explained as diffraction patterns of the waves. However, so established was Newton's physics of macroscopic bodies and so revered was he as an eminent scientist that belief in his notions of the nature of light held back the transition from the particle theory to the wave theory of light by about 100 years.

When the wave theory of light was finally accepted because of the preponderance of evidence that it was indeed a wave, it was then necessary to invent some sort of a medium to carry the undulations, i.e. something has to wave.

One cannot consider a water wave, for instance, if there were no water. But energy may be transmitted through water that has no net forward or backward motion by a traveling wave. Although there is no net transfer of water from one location to another, except up and down, until the wave is distorted by the shallowness of the water, there is the transverse transfer of a great amount of energy. Witness the tsunami.

Therefore, if light is a wave, interstellar space must consist of some medium capable of transmitting the light energy from place to place, just as the existence of an ocean is necessary to carry the energy of a water wave from one continent to another. Therefore, the word "ether" was used to identify the substratum in interstellar space that carried the energy of the light waves from the stars to the observer.

Light as a wave in the interstellar ether was the accepted truth until it was disproved by the famed Michaelson and Morley experiment. [Albert Abraham Michaelson (1852-1931) and Edward William Morley (1838-1923)]. If interstellar space consisted of substratum that carried light energy, the earth itself must be traveling through the ether, Michaelson and Morley measured the velocity of light at a given location on the earth at a given time, so that the point of measurement was moving with the rotation of the earth on its axis in the same direction as the revolution of the earth around the sun. Therefore, the velocity of the point of measurement through the ether would be the sum of the speed of the measurement point on the earth as it rotated daily on its axis (4.65×10^4 cm/sec), plus the rotational velocity of the earth as it traveled around the sun (4.75×10^5 cm/sec) or 5.2×10^5 cm/sec.

Six months later, when the earth was moving around the sun in the opposite direction, the measurement was again made when the measurement point on the earth was traveling around the earth in the same direction as the earth was moving around the sun. Therefore, the sum of the net speeds of that point on the earth, taken six months apart, (1.04×10^6 cm/sec) would be the total difference in the speed of that point through the ether as a result of the two measurements. The equipment used by Michaelson and Morley was accurate enough to detect a difference in the speed of light (3×10^{10} cm/sec) between the two measurement points if indeed the earth were embedded in the substratum of space. However, the experiment showed no measureable difference in the speed of light between the two observations. Thus, the existence of some ethereal substratum of which interstellar space was composed was demonstrated to be a false assumption.

The progress of the theory of the nature of light from that point on is a bit complicated, but it is now considered to be composed of particles called quanta or photons, that behave with wave-like properties. The wave-like properties are similar to the number of molecules of water along a water wave. The probability of finding photons at a given location on the path of travel of a light beam has the properties of sine wave; high probability at the anti-nodes and low probability at the nodes of the sinusoidal distribution of the photons.

A word about the techniques for measuring the speed of light would be in order at this point. In order to make a direct measurement of the speed of a particle it is necessary to identify the particle at point A and time A, and then to identify the particle again at point B and time B. The distance from A to B divided by the transit time -- miles per hour, or centimeters per second, for instance -- would be the speed of the particle between the two points. However, there is a fundamental difficulty in directly measuring the speed of a quantum (or photon) of light. When a photon of light is observed it is fully captured by the measuring equipment. No two measurements have ever been made on the same photon. So direct measurement of the speed of light is an impossibility.

However, indirect measurements supposedly can be made, but there are very serious problems with the indirect measurements. All indirect measurements of the speed of light are related to a mathematical equation developed by James Clerk Maxwell (1831-1929), who developed the electromagnetic theory of light. The electromagnetic theory is a mathematical description of the wave-like properties of light. Maxwell's equation relates the time rate of change of the electric field component of the electromagnetic wave to the space rate of change of the magnetic field component of the electromagnetic wave. These two components of the electromagnetic field are shown to be proportional to each other. The proportion can be made into an equation by adding a proportionality constant. The units of this constant, designated as "c", are a distance divided by a time -- in other words, a velocity. Therefore, it was concluded that the value of this proportionality constant represented the value of the speed of light. The value of the constant is 3×10^{10} cm/sec. Whenever someone makes an indirect measurement of the speed of light what he is really measuring is the value of that proportionality constant -- and it is always the same! Thus was born the conclusion that the velocity of light is independent of the velocity of the source and also independent of the velocity of the detector. P.W. Bridgeman, professor of Mathematics and Natural History in Harvard University, had this to say about the nature of light in his *The Logic of Modern Physics*, (The Macmillan Company, 1948):

"The most elementary examination of what light means in terms of direct experience shows that we never experience light itself, but our experience deals only with things lighted. This fundamental fact is never modified by the most complicated or refined physical experiments that have ever been devised; from the point of view of operations, light means nothing more than *things lighted*."

In spite of these overwhelming difficulties with the objectivity and the reality of the universe as described by modern science, and in light of the progression from the earliest recorded history of scientific belief in one erroneous concept after another, the scientists of today, as all those that have gone before, believe that they are finally on the right track to objective observation and description of universal truth that stands independent of man, his measurement techniques and the social environment that shape his perception and the interpretation of natural phenomena.

The Tenuous Nature of the Theory of Relativity

There is no question that the currently accepted theories of science will follow into the dustbin of history all those that held sway in earlier ages. The scientists of today loudly and emphatically affirm that they are on the track to absolute truth which stands independent of the observer and of the philosophy of science of which it is a part. Nothing could be farther from the truth! Cracks in the foundation holding up the theory of a forever expanding universe are already beginning to appear. Dr. Emil Wolf, of the Department of Physics and Astronomy, University of Rochester, New York published a paper in *Physical Review Letters*, 63, 20, 13 Nov. 1989, entitled *"Correlation-Induced Doppler-Like Frequency Shifts of Spectral Lines"*, in which he establishes mathematical justification for a scattering medium in interstellar space that adequately accounts for the observed Doppler effects from space radiation without invoking an expanding universe. Thus, an intellectually satisfying stable universe may be deduced that will adequately explain all the phenomena that up to now have required Einstein's problematical theory of relativity to explain.

Again, Einstein's theory of relativity has achieved the status of a *law* of physics, and is no longer in dispute in the universities. However, the theory of relativity has no basis in reality, but is a very useful mathematical construct by which accurate predictions of space, time and velocity near the speed of light (with respect to the observer) can be made. Because of the accuracy of the theory in explaining motion at velocities approaching the currently accepted notion of the velocity of light, it is assumed by many scientists to be a description of the real universe. The fallaciousness of such a belief is immediately obvious.

A case in point: Take two identical space ships. Placed side by side on the ground they are indistinguishable from each other. Let's have the two space ships travel in opposite directions from each other, each traveling at nine tenths the speed of light with respect to a given starting point in space. The question is asked, what is the velocity of one space ship with respect to the other? It cannot be 1.8c, since according to the theory of relativity, no signal can be propagated at a speed faster than the speed of light. According to the theory of relativity, as an object is accelerated to near the speed of light, the accelerating energy is converted into increasing the mass of the object as opposed to increasing its speed. Therefore if the speed of spaceship B is measured by observers in spaceship A they will find that space ship B is traveling nearer to the speed of light than 0.9c, but can never reach the speed of light. The effects of motion near the speed of light with respect to the observer of the object are that the object is heavier than when at rest with respect to the observer, its dimension in the direction of motion is shortened and its clocks are running slower than the clocks of the observer. Both mass, length and time in the direction of motion are relative quantities that are proportional to the speed of recession with respect to the observer.

That may be all well and good until we reverse the process and measure the same variables of spaceship A by an observer in spaceship B. The observer in spaceship B finds that the mass of spaceship A is greater than that of spaceship B, its clocks are running slower and A is shorter than B. Of course the mass, space and time as measured by each observer are assumed to be properties of the objects being measured, but they are *perceived* properties of the relationship between the observer and the object being measured -- hence the name of this theory as the theory of relativity. Both sets of observations cannot be fundamentally true. However, they are useful constructs in predicting perceived properties of matter at velocities approaching the current notion of the velocity of light.

THE THEORY OF EVOLUTION

Now let us turn to a brief examination of the theory of evolution. First, a little background. Up until about 1830 the general scientific belief concerning the history of the earth was that at one time it had existed as molten rock, and that during the cooling down process great cataclysmic events occurred, including sudden violent uplifts of mountains and major volcanic events. Evidence also existed for the impact of large meteoric bodies that caused major changes in the surface of the earth.

Scientific religionists tried to relate cataclysmic events to biblical accounts, notably the universal flood in 2348 BC. The common belief, even among many scientists of the time, was that the earth's history agreed with the biblical account of creation. The interpretation of the biblical account of creation by most of the scientists and theologians of the time was that the earth was created in a six day period about 4000 BC.

Catastrophism, as this school of thought came to be called, was attacked by the British Lawyer-turned-geologist, Charles Lyell (1797-1875). He expanded on the ideas formulated by a Scottish farmer, James Hutton, some fifty years earlier, who argued that the earth was not changed by sudden catastrophic events, but by unperceptively slow changes, such as those that are observed today. These would include the gradual uplift of mountains from the molten core of the earth and the rain-erosion of mountains.

These very gradual uplift and erosion actions would be repeated and continued over very long periods of time, resulting in a sort of perpetual motion machine passing through regular cycles of wearing away and rebuilding that finally made the planet suitable for the development of mankind. Lyell's version of geology came to be known as uniformitarianism, because of his fierce insistence that the processes that alter the earth are uniform over time.

Charles Darwin (1809-1882) built on the ideas of Lyell. He gathered data and developed his theory for 20 years before publishing his well-known book in 1859, *The Origin of Species by Means of Natural Selection*, or *The Preservation of Favoured Races in the Struggle for Life*.

Darwin's theory of evolution can be summarized in four statements:

1. Variation exists among individuals within species.
2. Organisms produce more offspring than the environment can support.
3. Competition exists among individuals.
4. The organisms whose variations best fit them to the environment are the ones who are most likely to survive, reproduce, and pass those desirable variations on to the next generation.

The confirmation of Darwin's theory was established through the subsequent study of paleontology, which defined a relative time element associated with the skeletal and fossil remains of plant and animal life. Simpler forms of life were generally found in deeper strata of sedimentary rock. The dating of the various times when different species existed came with the discovery of radioactivity in the 19th century.

There are two principal causes for concern in accepting the theory of evolution. One is the establishment of dates for prehistoric events by radioactive decay in rocks associated with the prehistoric sites. The other is the notion that animal life gradually progressed from the simplest single-cell forms to the eventual appearance and development of the homosapien species.

Axioms of the Theory of Evolution

Since all scientific disciplines begin with certain *a priori* assumptions that are not to be questioned or subjected to proof, it would be well to examine some of the *a priori* assumptions of the theory of evolution. These assumptions are based on the concept that the earth has had a benign history; that the composition and status of the earth has always been as it is today and that the natural aging processes of mountain uplift and erosion have always occurred under the same conditions and at the same rate as they do today.

Geological Dating Techniques

Radioactive isotopes of given elements decay in exponential form at very precise rates. One expression of the exponential rate of decay is that the same fraction of radioactive material decays to the daughter isotope over the same period of time. Thus, for instance, half of the remaining amount of any given sample of radioactive material will decay over the same period of time. This time is called the half-life of the material.

The rate of decay varies with the chemical identity of the element. The decay rates of the various radioactive elements vary from extremely long to extremely short times. The half-life of the radioactive isotope of uranium-238, for instance, is 4.5×10^9 years, whereas the half-life of polonium-215 is 1.8×10^{-3} seconds.

Carbon-14 Dating

An important isotope used in geological dating of plant material is the radioactive isotope Carbon-14, which was discovered February 27, 1940, by Martin Kamen and Sam Ruben. C14 is produced in the upper layers of the troposphere and the stratosphere by thermal neutrons absorbed by nitrogen atoms. Carbon-14 reacts with oxygen in the atmosphere to produce carbon dioxide, which is an integral component of all plant life. C14 has a half-life of 5730 years, and has a relative concentration of up to only one part per billion of all the naturally-occurring carbon on the earth. Present techniques for the measurement of the fraction of C14 in a specimen are such that concentrations representing about 10 half lives are at the limit of detection. That limit would represent about 60,000 years. (Minze Stuiver and Henry A. Polach, "Reporting of C14 Data," *Radiocarbon,* Vol. 19, No. 3, 1977, pp 355-363).

The Proportion of C14 in the Atmosphere

We will begin with the relative concentration in the atmosphere of C14. C14 is produced in the upper atmosphere by cosmic ray bombardment of the earth from interstellar space. One *a priori* assumption is that the cosmic ray bombardment intensity has remained constant over time. A second *a priori* assumption is that there have been no other sources of C14 in the atmosphere. A third *a priori* assumption is that the relative concentration of C14 in the atmosphere has remained constant over time.

Evidence that none of these assumptions is true abounds. Yet, characteristic of the proponents of all the scientific theories of the past, those who believe that the theory of evolution is no longer a theory, but a law of science, ignore the contrary data. One case in point: There was a baby mammoth discovered in 1977 that was subjected to radiocarbon dating. Examination of one part of the carcass revealed that the animal lived 40,000 years ago.

Examination of another part revealed that the animal lived 26,000 years ago, and wood found in the immediate vicinity of the carcass was shown by radiocarbon dating to be from 9,000 to 10,000 years old.

What are some of the things that could change the concentration of C14 in the atmosphere? Insolation. There are periodically great fluctuations in the intensity of radiation from the sun. Sunburst activity, for instance, can be intense enough at times to interrupt radio communication all over the earth. The Aurora Borealis is caused by the creation of ions in the atmosphere from solar radiation. The radiation influx has varied so much, for instance, that there was an occurrence in the 1980s in which the intensity of the Aurora Borealis, normally visible only in the polar-regions, was so great that it was observed as far south as the State of Georgia.

The amount of cosmic rays penetrating the earth's atmosphere is also affected by the earth's magnetic field which deflects cosmic rays. Precise measurements taken over the last 140 years have shown a steady decay in the strength of the earth's magnetic field. This means that there has been a steady increase in radiocarbon production (which would increase the ratio of C14 in the atmosphere). Additional variations in the atmospheric content of C14 undoubtedly result from volcanic eruptions, which are known to contain great quantities of carbon and carbon dioxide that are devoid of C14. Since the ratio of C14 to C12 (the stable isotope of carbon) is only on the order of one part per billion, a relatively small fraction of C14 could alter that ratio in a major way.

The Inaccuracy of C14 dating

The gross inaccuracy of C14 dating has been prominently reported in the scientific literature, although apparently ignored by the proponents of uniformitarianism. To cite just two examples:

From the *Creation Research Society Quarterly*,

"For the past century and a half, careful measurements of the earth's magnetic field have been conducted. The rapid decay of the magnetic field is startling. Assuming that this rate is constant, scientists are able to show mathematically how strong the magnetic field was in the distant past. Instead of an age in the millions or billions of years, however, the magnetic field can be projected back in time less than 20,000 years.
"The world could not exist with the powerful magnetic field projected beyond 20,000 years. This finding is strong evidence for a young earth." (*CRSQ*, 1972, 9:1, p. 47).

From the Reader's Digest,

"More trouble appeared several years ago with studies of bristlecone pine borings.

"These trees are considered by most scientists as the oldest living matter known on earth. However, C14 tests made with wood from these pines of at least approximately known age showed that C14 readings were in error from a few centuries up to a thousand years. This find cast further doubt on the assumptions of the method." (The Reader's Digest, 12/1972. pp. 86-90).

Potassium-Argon Dating

One of the chief problems associated with dating animal remains by radioactive decay is the assumption that the radioisotope was generated at the same time that the animal lived, i.e., that it was a pure sample. If that element had been present at the site before the animal lived, part of it would have already decayed which would give the time of the site a much older date than would the assumption of a pure sample.

Potassium-Argon dating is the only viable technique for dating very old archaeological materials. Geologists have used this method to establish dates for rocks which give results as much as 4 billion years old. It is based on the fact that one of the radioactive isotopes of Potassium -- Potassium 40(K40), decays to the gas Argon as Argon 40 (Ar40). By comparing the proportion of K40 to Ar40 in a sample of volcanic rock, and knowing the decay rate of K40, the date that the rock was formed is determined.

Potassium is one of the most abundant elements in the Earth's crust (2.4% by mass). One out of every 10,000 Potassium atoms is the radioactive isotope K40. When rocks are heated to the melting point, any Ar40 contained in them is released into the atmosphere. When the rock re-crystallizes it becomes impermeable to gasses again. As the K40 in the rock decays into Ar40, the gas is trapped in the rock.

So when a rock is tested for Ar40 the ratio of Ar40 to K40 would show the time since the rock was molten, not necessarily the time when it was associated with some prehistoric animal found in association with the rock. Unless the animal were killed at the time the rock solidified the Ar40 time clock would show a time earlier than when it was associated with the animal being evaluated. Other serious difficulties exist with this method of dating which appear to make it completely unreliable as a geological time clock. The testing performed on volcanic rocks in Hawaii should be sufficient to exclude this method from further use. Ar40 tests performed on rocks from lava flows in Hawaii known to have occurred in 1800 and 1801 were dated by the radioactive potassium-argon method as having occurred 160 million years ago and 2.96 billion years ago! And yet the method is still in use! The conclusion of those who made the tests was that the samples of magma that were extracted from the lava flow for testing had to have been from intrusion rocks rather than a part of the lava flow, and the data was ignored! This blind adherence to an obviously flawed dating method is not lost on other members of the scientific community.

From the *Research Communications Network Report,*

"Dr. Robert V. Gentry is the world's leading authority on radio-halo research. He has published a remarkable series of papers in such distinguished journals as *Nature, Science, and Annual Review of Nuclear Science*. His findings are of great significance to the question of radiometric dating. Among his carefully drawn conclusions are the following: Earth's primordial crustal rocks, rather than cooling and solidifying over millions or billions of years, crystallized almost instantaneously. Some geological formations thought to be 100 million years old are in reality only several thousand years old." (*Research Communications Network Report*, 2/10/1977, p. 3).

UNIFORMITARIANISM

The principle questionable assumption of the theory of evolution is the assumption of a benign history of the earth -- that the processes of erosion and mountain uplift have always occurred at the same rate that they are observed to occur today. This is the basic premise that separates the concept of uniformitarianism from the concept of catastrophism. The evolutionists claim that there is no viable evidence to counter the theory of evolution, yet such evidence is overwhelming. However, this evidence has not been acknowledged by the proponents of uniformitarianism for the very same reasons that have held back the demise of all the incorrect theories of the past. Because of such dogmatism it can be said that *the universities, that are the champions of the current theories, have done more than any other institution to hold back the advancement of science!*

To support that statement we cite the case of the contemporary scientist, Dr. Immanuel Velikovsky (1895-1979). Because of the importance of his controversial contribution to the subject at hand, it is prudent to include here a brief biography of his life.

On June 10, 1895, Immanuel Velikovsky was born in Vitebsk, Russia. He learned several languages as a child, performed exceptionally well in Russian and mathematics at the Medvednikov Gymnasium after moving to Moscow, and graduated with a gold medal in 1913. He then traveled to Europe, visiting Palestine, briefly studying medicine at

Montpelier, France, and taking premedical courses at the University of Edinburgh. Having returned to Russia before the outbreak of World War I, Velikovsky enrolled in the University of Moscow, and received a medical degree in 1921. Then he left Russia for Berlin, where he edited the *Scripta Universitatis*, for which Albert Einstein prepared the mathematical-physical section. Velikovsky's work on the *Scripta* ultimately led to the birth of the Hebrew University in Jerusalem. From 1924 to 1939 Velikovsky lived in Palestine, practicing psycho-analysis -- he had studied under Freud's pupil, Wilhelm Stekel, in Vienna -- and editing *Scripta Academica Hierosolymitana.*

In 1939 Velikovsky took a sabbatical year, traveling with his family to New York only a few weeks before World War II tore Europe apart. In April, 1940, Velikovsky was first struck by the idea that a great natural catastrophe had taken place at the time of the Israelites' Exodus from Egypt -- a time when plagues occurred, the Sea of Passage parted, Mt. Sinai erupted, and the pillar of cloud and fire moved in the sky. Velikovsky wondered: Does any Egyptian record of a similar catastrophe exist? He found the answer in an obscure papyrus stored in Leiden, Holland --the lamentations of an Egyptian sage, Ipuwer. The Ipuwer document, Velikovsky became convinced, parallels the Book of Exodus, describing the same natural catastrophes and the same plagues. As a result he began to reconstruct ancient Middle Eastern history, taking this catastrophe -- which brought the downfall of the Egyptian Middle Kingdom -- as a starting point from which to synchronize the histories of Egypt and Israel. He titled his work *Ages in Chaos*. The cause of the catastrophe terminating the Middle Kingdom remained unexplained.

One afternoon in October, 1940, Velikovsky noticed an important fact: the Book of Joshua describes a destructive shower of meteorites occurring before the sun "stood still" in the sky (Joshua 10:11). Could this be a coincidence, or were the ancients recording a cosmic disturbance that must have shaken the entire Earth and might have been related to the upheavals approximately 50 years earlier during the Exodus? A survey of other sources around the world convinced Velikovsky that a global cataclysm had indeed overtaken the Earth, and that Venus played a decisive role in that cataclysm. For 10 years he researched and wrote *Ages in Chaos* and *Worlds in Collision*.

Even before its appearance, the book was enveloped in furious controversy. We quote from an article by David Stove published in *Pensee*, May 1972, entitled *"The Scientific Mafia"*.

"The 'professional scientists' campaign against *Worlds in Collision* began well before the book appeared. Harlow Shapley, probably the best-known American astronomer alive today, led an energetic attempt to stop the publisher, MacMillan, from publishing the book. He arranged for denunciations of the book, still before its appearance, by an astronomer, a geologist and an archeologist, in a learned journal. None of them had read the book. When it did appear, denunciatory reviews were arranged, again in several instances, by professors who boasted of never having read the book.

"Velikovsky was rigorously excluded from access to learned journals for his replies. Then Shapley and others really got busy on the old-boy circuit. They forced the firing of the Senior Editor of MacMillan responsible for accepting the Velikovsky manuscript. (He had been with the firm 25 years). They forced the firing of the director of the famous Hayden Planetarium in New York, because he proposed to take Velikovsky seriously enough to mount a display about the theory.

"The MacMillan representatives all over the country began to report that science professors in the universities were refusing to see them. MacMillan finally caved in, and prevailed on Velikovsky to let them transfer their best-selling property to a competitor, Doubleday, which, as it has no textbook division, is not susceptible of professorial blackmail."

Since Velikovsky's work was at variance with the accepted scientific notions of the day, the treatment that he received from the scientific community was not different from the treatment by society of Socrates, Galileo, and others who dared to challenge the "truths" of their age.

CATASTROPHISM

The Velikovsky Findings

Among the conclusions of Velikovsky's work that have so disturbed the proponents of uniformitarianism are the following:

The Erratic Venusian Orbit

The sun-earth-moon system has undergone two periods of extreme disturbance which have changed the parameters of the system by large amounts within the past few thousand years. In the first period, the earth had two near-collisions with Venus, the first being in the time of the Hebrew exodus from Egypt and the second being on the day during Joshua's campaigns when the sun and moon stood still. These near-collisions were about 50 years apart, and they took place sometime around 1500 BC. Velikovsky points to literature of the period that refers to the 'horns' of Venus, an obvious reference to the crescent planet, which came near enough to the earth to be seen as a sphere rather than as a point of light, and since it is nearer to the sun than is the earth, would be seen in phases, as is the moon.

The Erratic Martian Orbit

During the second period, the earth had two or more near-collisions with Mars, and Velikovsky gives exact dates for the first and last of them. The first was on February 26, 746 BC and the last was on Mar 23, 686 BC. Since March 23, 686 BC the parameters of the sun, moon, and planets have been undisturbed. For at least part of the time between 746 and 686 BC, the month had 36 days while the year continued to have 360 days, according to Velikovsky. Thus he explains the fact that the oldest known Roman calendar had only 10 named months.

{Author's note: The names of the last four of the ten months of the year were September, October, November and December --Seventh, Eighth, Ninth and Tenth. Thus, sometime before 686 BC the continued disruptions of the lunar orbit by near-passes of the planet Mars resulted in a change of the orbital period from 36 to 29.53 days. This caused so much trouble with the calendar of 36-day months that in 45 BC Julius Caesar commissioned the 12-month calendar to be developed. To the 10-month calendar they added in the middle of the year two new months-July and August (Julius and Augustus). This Julian calendar held sway until 1582, when Pope Gregory XIII instituted a 'new and improved' calendar (called the Gregorian calendar) that replaced the older Julian calendar. The new calendar was quickly adopted in most Catholic countries, but the many Protestant and Orthodox countries continued to use the older Julian calendar for centuries. Russia, for instance, continued to use the Julian calendar until Feb 14, 1918.}

Velikovsky also predicted that Venus had retrograde rotation, a heavy hydrocarbon atmosphere and a surface temperature on the order of 800^0 C. All of those predictions were verified by subsequent space probes of Venus conducted by NASA.

Acknowledgment of Velikovsky's Achievements

Although Velikovsky was generally treated as a modern-day Galileo and was ostracized by the scientific community, yet there were a few, who, although disagreeing with his theories, nevertheless respected him as a scientist and took his work seriously. Among them was Dr. H.H. Hess, Chairman of the Space Board of the National Academy of Science, who wrote to Dr. Velikovsky the following letter:

"March 15, 1963
Dear Velikovsky:
We are philosophically miles apart because basically we do not accept each other's form of reasoning -- logic. I am of course quite convinced of your sincerity and I also admire the vast fund of information which you have painstakingly acquired over the years.
I am not about to be converted to your form of reasoning though it certainly has had successes. You have

after all predicted that Jupiter would be a source of radio noise, that Venus would have a high surface temperature, that the sun and bodies of the solar system would have large electrical charges and several other such predictions. Some of these predictions were said to be impossible when you made them. All of them were predicted long before proof that they were correct came to hand. Conversely I do not know of any specific prediction you made that has since been proven to be false. I suspect the merit lies in that you have a good basic background in the natural sciences and you are quite uninhibited by the prejudices and probability taboos which confine the thinking of most of us.

Whether you are right or wrong I believe you deserve a fair hearing.

Kindest regards,
(signed) H.H. Hess"

Apart from Velikovsky, there is incontrovertible evidence that the earth has had a very stormy rather than a benign history. Even the evolutionists invoke catastrophism to account for the rather sudden demise of the dinosaurs, suggesting that they may have been extinguished by the impact of a large asteroid. In an article from the Department of Paleobiology of the Smithsonian National Museum of Natural History, available on their website, a thesis is developed that 65 million years ago an asteroid on the order of 100 miles wide impacted the Yucatan coastline at an angle from the southeast. The heat of impact sent a searing vapor cloud speeding northward which within minutes, set the North American continent aflame. The fireball and the darkness that followed caused major plant extinctions in North America. Environmental consequences led to the global extinction of many plants and animals, including the dinosaurs.

Meteoric Impact

All the inner bodies in our solar system have been heavily bombarded by meteorites throughout their history. The surfaces of the Moon, Mars and Mercury, where other geologic processes stopped millions of years ago, record this bombardment clearly. On the Earth, however, which has been even more heavily impacted than the Moon, craters are continually erased by erosion and re-deposition as well as by volcanic resurfacing and tectonic activity. Thus only about 120 terrestrial impact craters have been recognized, the majority in geologically stable regions of North America, Europe and Australia where most exploration has taken place. Spacecraft orbital imagery has helped to identify meteoric structures in more remote locations.

Shifts of the Earth's Axis

There is also compelling evidence that the earth has shifted on its axis numerous times. Such shifts would be followed by catastrophic climatic changes. Evidence of shifts of the axis of the earth are deduced from observed changes in the magnetic orientation of sedimentation and of volcanic floe. Ferromagnetic material deposited as sedimentation on the ocean floor would align itself with the magnetic field of earth, and once deposited would remain fixed in orientation. Likewise, the ferromagnetic material in lava flows as they cooled toward the solid state would tend to orient themselves according to the earth's magnetic field. Once solidified, they would then be immobile. Examination of samples from such lava flows have shown that the earth has been reversed on its axis numbers of times. Attempts to date such catastrophic shifts in the earth's orientation have been made using the radioactive decay rates of the potassium-argon system. However, as explained earlier, this system may produce highly erroneous results because of problems with the *a priori* assumptions and other limitations associated with the method. As a commanding case in point, as mentioned earlier, volcanic rocks produced by Hawaiian lava flows in 1800 and 1801 were dated by the radioactive potassium-argon method as having occurred 160 million and 2.96 billion years ago!

An account was published in Time Magazine in 1946 of a core taken from the abysmal depths of the Atlantic Ocean along which the ferromagnetic orientation was measured. That test revealed that the axis of the earth had been reversed 167 times, and that the last reversal was within historic times! (It has recently been discovered that the sun undergoes a magnetic reversal every 11 years at the peak of each solar cycle). (*Science News*, Dec. 6, 2003; Vol. 64, No. 23, p. 355).

The Formation of the Grand Canyon

From the *Hillman Wonders of the World*, "[The Grand Canyon] was slow in making. It took the erosive effects of the Colorado River millions of years to carve a deep swath through billion-plus year-old rock layers." It is interesting to cogitate on how that could have occurred with the canyon being up to 18 miles wide from rim to rim, and with the north rim of the canyon in places being 1,000 feet higher than the south rim and the canyon being up to a mile deep! One wonders why the other four major rivers in Utah (curiously enough, all heading within 100 miles of one another in the high Uinta Mountains) flowing to the north, south and east parts of the State have not approximated the same effects as the Colorado River (which occurred along only a part of its length). Again, the theory of evolution must invoke catastrophic events to provide the geologic rift called the Grand Canyon.

The National Park Service, U.S. Department of the Interior, published this description of the formation of the Grand Canyon:

> "The Grand Canyon is an erosional feature that owes its existence to the Colorado River. Of equal importance are the forces of erosion that have shaped and continue to shape the canyon today.

"These include running water from rain, snowmelt, and tributary streams which enter the canyon throughout its length.

"The climate at the Grand Canyon is classified as semi-arid. The South Rim receives 15 inches/*38 cm* of precipitation each year. The bottom of the canyon receives 8 inches/20 cm. The rain comes suddenly in violent storms, particularly in the late summer of each year. The power of erosion is therefore more evident here than in other places which receive more rain."

The theory of evolution recognizes the influence of continental uplift from tectonic plate movement as the source of the rift that created the Grand Canyon, but insists that the changes occurred at imperceptible rates over millions of years.

The Erosion Rate of Niagara Falls

The following is an excerpt from pages 81-84 of *In the Minds of Men* (1987) by Ian Taylor:

"Lyell talked to a local inhabitant and was told that the falls retreat about three feet a year. He assumed that this was an exaggerated claim and concluded that one foot a year would be a more likely figure

(Lyell 1867, 1:361). On the basis of this guess, it was then a simple matter to equate 35,000 feet, or seven miles, as 35,000 years that the falls had taken to cut the gorge from the escarpment to the place it occupied in the year of his visit, which is how he arrived at the figure that he announced to the scientific world. The principle was sound enough, but his method can hardly be called scientific or even honest. (Bailey 1962, 149).

"In recent years the estimate base has been revised downward, but in the mid-nineteenth century it had a most significant impact on the common man's beliefs. Lyell's *Principles of Geology*, as already mentioned, was published in 1830-33, and although it was met with opposition at first, it eventually became the standard work on the subject for the next fifty years, running to twelve editions. Charles Lyell became Sir Charles in 1848, principally because of his Scottish land holdings. To the Victorian mind, this title gave his name and books tremendous credibility and authority; in a similar way today, the news media seek out a scientist with a legitimate Ph.D. when they want an authoritative scientific opinion. Lyell's figure of 35,000 years for the cutting of the Niagara gorge was thus accepted as an actual measurement made by a gentleman of integrity and quite beyond dispute. For the next few generations this estimate served wonderfully to demolish any credence in Archbishop Ussher's date of creation and made the attempt to finish once and for all the orthodox belief in the Mosaic Flood, which was alleged to have occurred a mere four-and-half thousand years ago.

"Measurement of the rate of recession of Niagara Falls has been made periodically since 1841, the date of Lyell 's visit, and these published figures show that, far from exaggerating, the local inhabitant was too conservative. A rate of four or five feet a year is closer to the facts. (Tovell 1979, 16). Assuming, as Lyell did, that the rate of recession had always been the same, this measured value reduces the age of the falls to between seven and nine thousand years. Had it been honestly reported in the first place, this would not have been regarded as refutation but rather a near confirmation of the Mosaic Flood!"

Animals from Temperate Climates Found Frozen in Arctic Regions

From *In the Minds of Men,* by Ian T. Taylor,

"The year 1901 provided a unique opportunity to make a firsthand scientific study of a mammoth that had then recently been exposed on the banks of the Beresovka River in northeastern Siberia and sixty miles inside the Arctic Circle (Digby 1926; Billow 1981; Pfizenmayer 1939; Sanderson 1960). The mammoth was found frozen in a sitting position in what is technically referred to as muck and located in the middle of an ancient landslide. The flesh and even the eyeballs were so well preserved that the expedition's sled dogs had plenty of fresh meat to eat. Death must have come to this specimen very quickly, because the blood still contained some oxygen and was preserved sufficiently well to establish the relationship to the blood of today's Indian elephant, although distinct anatomical differences would not necessarily classify them as the same species. There was well-preserved food in the mouth and twenty-four pounds of un-dissolved and identifiable plants in the stomach....Now all these details were soberly reported in the scientific journals of the day, including the annual report of the Smithsonian Institution for 1903, yet geology textbooks still insist on the uniformitarian explanation that the unfortunate creature-- as if it were the only one --must have stumbled and died where it fell amid the snow-covered wasteland. Imaginative paintings of the woolly mammoth by artists such as Burian have traditionally shown these animals in a winter landscape. Yet the reported examination of the skin showed that the creature had no sebaceous glands and therefore could not have oiled its fur to survive in Arctic conditions.

Further, more than fifty varieties of herbs, grasses and mosses, some of which only grow in temperate climates, were identified in the stomach. (Dillow 1981, 371-80). The buttercups, for example, were in seed and fixed the time of death in late July or early August.

"All these and many more details of the mammoth have been available in English to anyone willing to make inquiry at even a modest university library. Yet for most who actually do so and who popularize the mammoth mystery, there is the natural tendency to develop a theory to explain the mystery. Unfortunately, details that do not fit the particular theory often go unreported. Since 1981, Joseph Dillow's *The Waters Above* has provided a popular explanation for the Beresovka Mammoth. His explanation relies upon a catastrophic drop in atmospheric temperature causing the creature to literally freeze in its tracks. The evidence cited for this is the preserved remains of the delicate sedges and grasses found in the stomach. However, unknown to Dillow was the fact that, like the elephant, the mammoth had two stomachs; the first was simply a holding vessel while the second was for the actual digestion. The scientists who had examined the Beresovka mammoth reported the contents of the first stomach. Prior to Dillow's work there were others who had drawn a similar conclusion based upon the erroneous understanding that the mammoth contained only a single stomach. Their theory proposed that the earth had passed through the icy tail of a comet. The ice particles at the temperature of outer space were caught up in the earth's gravitational and magnetic fields and dumped on the magnetic north and south poles. (Gow 1972; Patten 1976; Sears 1979). It was argued that there could have subsequently been some glaciation at the edges of the instantly formed ice field. The ice-dump theory was claimed to account for Canada's permafrost, ice caves between lava rocks (Patten 1976, 120), and the Ross sea-bed core evidence that indicates that Antarctica only became ice-covered as recently as six thousand years ago. The bottom line is that thousands of these mammoths had lived in the Arctic when the climate was moderate and there was sufficient food. There was an Ice-Age in which most of the mammoths died and decayed and a very few remained frozen. Beyond that no one knows exactly what or how it happened." (*In the Minds of Men, Darwin and the New World Order*, Ian T. Taylor, Fifth Edition 2003.Chapter 4).

The above accounts represent only a few of the scientific evidences that the earth has not had a benign, unperturbed history. As mentioned earlier, evidence for a very cataclysmic history, extending into recorded historical times, abounds.

REVEALED RELIGION

Religious Methodology vs. Scientific Methodology

The formalism of the scientific method requires that the concepts addressed be in some way measurable, repeatable and communicable. Before a scientific theory can be validated, it must be tested to prove its validity. The testing is generally directed to demonstrate that the theory is not true or valid. If it cannot be proven to be invalid, then it is accepted.

Thus the scientific method restricts itself to concepts that are at least theoretically measurable in the laboratory. The nature of science is such that it cannot address subjects of religion since matters of the conscience cannot be subjected to controlled experimental procedures. Science does not rule out religion, it is simply not structured to address religious concepts.

By using the term "revealed religion" we restrict ourselves to religious concepts that are announced in the scriptures and those that have been announced by the prophets of God that have not been included in the scriptures. It is recognized that there are many interpretations of that which is written in the scriptures. However, it is also understood that there are many who interpret the scriptures to fit and bolster preconceived notions. Witness the fact that the world-wide number of Christian denominations is over 9,000, with 635 registered in the United States alone, most of which have deduced their various doctrines from the same Bible. (World Christian Data Base, Center for the Study of Global Christianity, Gordon-Conwell Theological Seminary).

The Interpretation of Religious Literature

How do we differentiate our own interpretation of the scriptures from the many others? This is done by recognizing that the Church of Jesus Christ of Latter-day Saints represents the restoration through the prophet Joseph Smith of the original Church of Jesus Christ organized by the Savior Himself during his mortal ministry. The defining authority in this church are its ecclesiastical officers, who are called by God to their various positions of service, possessing the authority of the holy Melchizedek priesthood, and who speak as moved upon by the Holy Ghost.

Intrinsic to this restored Kingdom of God on the earth is the gift of the Holy Ghost, one of whose functions is to confirm eternal truths in the hearts of those possessed of this gift which is bestowed upon the believers by the laying on of hands of those possessing requisite priesthood authority following their baptism. Thus, the membership of the Mormon Church have the right to inspired confirmation of the declarations of the prophets.

In the discussion of religion *a priori* postulates are not addressed, since they are part of the formalism of science, not of religion. To gain knowledge by the religious process one starts with the concept of faith. As opposed to the scientific method, faith requires that a concept be assumed to be true *before* it is tested for its truthfulness. The assumption of the truthfulness of a concept with sufficient conviction to put it into practice is an act of faith. That assumption cannot be a mere postulate, but it must be a sincere acceptance of the concept as part of the structure of one's being. Thus faith could be said to be a motivating belief in that which is true accompanied by confirming, compliant action. (The Book of Mormon, Alma 32:2).

After faithful compliance with the principle in question, the Holy Ghost reveals to the individual the truthfulness of the doctrine that he is practicing. This revelation from a member of the Godhead is by definition true, since God cannot lie, otherwise He would cease to be God. (George Q. Cannon, *Gospel Truth: Discourses and Writings of President George Q. Cannon*, 1987 Deseret Book Company).

Revealed religion is generally directed to the revelation of those principles of human behavior designed to qualify one to be worthy to return to the presence of his Father in Heaven following his mortal experience. However, peripheral matters reveal information in the same realms that comprise the very themes of some of the scientific disciplines. These areas of overlap are in general

69

disagreement, and are the cause of much controversy. Thus the academician who is also a religionist is forced to make a choice between that which is taught in the universities to be the demonstrated truth of science and that which he learns in his religious curricula to be the revealed word of God.

It is perhaps surprising to note that revealed religion supports the notion of uniformitarianism in much of its doctrine, but also supports catastrophism in other particular areas. The uniformitarianism concepts relate to man's progress from a spirit child of God, born in the indeterminate past, to the possibility of progress through time and eternity to eventually be perfected in righteousness and achieve an eternal adulthood to become like his eternal Father. In this process the progress is gradual, consistent with the philosophy of uniformitarianism. An expression of this concept in the terminology of religion is found in the Book of Mormon, 2 Nephi 28:30, as follows:

> "For behold, thus saith the Lord God: I will give unto the children of men line upon line, precept upon precept, here a little and there a little; and blessed are those who hearken unto my precepts, and lend an ear unto my counsel, for they shall learn wisdom; for unto him that receiveth I will give more; and from them that shall say, We have enough, from them shall be taken away even that which they have."

The catastrophism concepts relate specifically to the history of the earth -- both with the concept of the earth as a planet, and with the concept of a timeline dating pre-historical events. These concepts are at sharp variance with the theory of evolution. If they are true concepts the theory of evolution must necessarily be false in those areas of conflict. On the other hand, those who profess that their information represents revealed religion are necessarily mistaken in those beliefs.

The Revealed History of Planet Earth: Evidence of Catastrophism Recorded in the Scriptures

The Universal Flood

The first significant catastrophic event recorded in the Bible is the account of the universal flood in the days of Noah. Noah was born in 2,948 BC and was 600 years old when the flood occurred in 2,348 BC. The rains that resulted in the flood lasted for 40 days and 40 nights; but not only was there a great deluge, but the *fountains of the great deep were broken up.* (Genesis 7:11. Italics ours). The entire earth was covered with water, and all living things except those that were in the ark with Noah, were destroyed. We learn in Genesis 7:20:

"Fifteen cubits upward did the waters prevail; and the mountains were covered."

That statement is not consistent with the configuration of the earth today, which causes many who do not have confidence in the truthfulness of the scriptures to believe that the scriptural account is a groundless fable. However, the topology of the earth in Noah's time was not as it is today. It was *after* Noah's time that a second significant catastrophic event is revealed in the Biblical record. This event is recorded in a single sentence without elaboration in the Book of Genesis:

"And unto Eber were born two sons: the name of one was Peleg; for in his days was the earth divided. (Genesis 10:25).

The Earth Divided

We are left to conjecture as to how the earth was divided. But prior to Peleg's time the earth consisted of the sea and a single land mass. So it is not without the realm of possibility that a 15-cubit increase in the water level could have covered the entire earth.

Some people have questioned the validity of the account of the flood on the grounds that it would have been impossible for Noah to have collected the animals from all the continents to go into the Ark. That objection to the reality of the universal flood is taken care of by an understanding from the same literary source of the flood that at the time of the flood the earth's land mass was in one piece.

The Sun Standing Still

There is also the account in the Book of Joshua of the Lord causing the earth to cease its rotation for about a whole day (Joshua 10:12-14), and an account in 2 Kings where the earth was rotated back by 10 degrees. (2 Kings 20:8-11).

These occurrences are considered as mere fables by the scientific community, and such objections are offered as; if the earth were to stop rotating for about a day or for only 40 minutes, the resulting 1,015 MPH wind at the equator, with correspondingly slower speeds at increasing latitudes -- decreasing to 500 MPH at latitude 60 degrees -- would sweep everything off the surface of the earth and destroy all life except in the vicinity of the polar regions.

Those conclusions, that accept for argument the supposition that God could stop or retard the rotation of the earth, do not admit the possibility that God could also stop the rotation of the air above the earth. They also assume that the action taken was a sudden action, which it may not have been. Since God is the creator of the universe, and has all knowledge – *"O how great the holiness of our God! For he knoweth all things, and there is not anything save he knows it."* (2 Nephi 9:20). -- He would certainly have the power to bring to pass anything that He would undertake to do. The objections of the scientific community to the miraculous works of God attempt to limit Him to the pitiful extent of their own knowledge of the universe.

Natural Catastrophes at the Time of the Crucifixion of the Savior

The scriptures record cataclysmic destruction at the time of the crucifixion, including cities destroyed by fire, sunk into the depths of the sea, buried in the depths of the earth and covered with floods. In addition, many other *great and terrible destructions* were caused to come upon the land. The Book of Mormon records that because of the great and terrible destructions at the time of the crucifixion --

> *"...the whole face of the land was changed*...And the highways were broken up, and the level roads were spoiled, and many smooth places became rough... And thus *the face of the whole earth became deformed,* because of the tempests, and the thunderings, and the lightnings, and the quaking of the earth. And behold, the rocks were rent in twain; they were broken up upon the face of the whole earth, insomuch that they were found in broken fragments, and in seams and in cracks, upon all the face of the land." (3 Nephi 8:12-18. Italics ours).

Driving east about 50 miles out of San Diego, California one comes to a region where large, fractured boulders cover the ground. There is also a rather extensive region in the province of San Juan, Argentina where similar fractured boulders cover the surface of the ground, perfectly consistent with the information recorded in the 8th chapter of 3 Nephi.

The True Creation Account

The Biblical account of the creation is extremely sketchy, and is subject to many interpretations, including the very literal interpretation that the earth was created in six 24-hour days in 4004 BC. Such superficial erroneous interpretations are justifiably denounced by the scientific community. However, when the sketchy biblical account of the creation is considered in light of the more complete accounts recorded in *The Pearl of Great Price* a much more logical account of the creation process is revealed. In Abraham, chapter 4, we learn that the scriptures record the account of two creations --first a spiritual creation and secondly a temporal creation. First, however, the heavens and the earth were formed,

> "And they went down at the beginning, and they, that is, the Gods, *organized and formed* the heavens and the earth....
>
> "And they (the Gods) said: Let there be light; and there was light....

"And the Gods ordered the expanse, so that it divided the waters which were under the expanse from the waters which were above the expanse; *and it was so even as they ordered....*

"And the Gods ordered, saying: Let the waters under the heaven be gathered together unto one place, and let the earth come up dry, *and it was so as they ordered.*" (Abraham 4:1,3,7,9. Italics ours).

All of the above is in the past tense -- "*They organized and formed*", "*it was so even as they ordered.*"

The Spiritual Creation

Beginning with Abraham 1:11 the Gods speak of what they plan to do,

"Let us prepare the earth to bring forth grass,"

The Gods organized the earth "*to bring forth grass.*" The earth was prepared for planting, but the planting had not yet taken place. Next, the Gods caused day and night to occur.

"And the Gods organized the lights in the expanse of the heaven, and caused them to divide the day from the night." (Abraham 4:14).

After this they prepared the earth for animal life.

76

"And the Gods said: Let us prepare the waters to bring forth abundantly the moving creatures that have life; and the fowl, that they may fly above the earth in the open expanse of heaven." (Abraham 4:20).

"And the *Gods prepared the waters....*

"And *the God's prepared the earth* to bring forth the living creature after his kind, cattle and creeping things, and beast of the earth after their kind....

"And the Gods organized the earth *to bring forth the beasts* after their kind, and cattle after their kind, and everything that creepeth upon the earth after its kind;" (Abraham 4:21, 24-25. Italics ours).

The Abrahamic narrative next reports the plans of the Gods to put man on the earth.

"*Let us go down* and form man in our image....

"So the Gods *went down to organize man* in their own image, in the image of the Gods to form they him, male and female *to form* they them....

"And the Gods said, *we will bless them...We will cause them* to be fruitful and multiply...*we will give them* every herb bearing seed...and also *we will give them* every green herb for meat, *and all these things shall be thus organized."* (Abraham 4:26-30. Italics ours).

All the above reports planned for future action, as confirmed in Abraham 5:5:

"...for the Gods had not caused it to rain upon the earth when they counseled to do them, and had not formed a man to till the ground."

The Temporal Creation

After organizing the earth with its dry land, sea, air and moisture, and then preparing it for plant and animal life, and for the advent of Adam, Adam was brought to the earth.

"And the Gods *formed man* from the dust of the ground." (Abraham 5:7. Italics ours).

Next, the Gods caused vegetation to grow on the earth.

"And the Gods *planted a garden*, eastward in Eden...And out of the ground *made the Gods to grow every tree* that is pleasant to the sight and good for food." (Abraham 5:8-9. Italics ours).

Next, the Gods brought the animals to the earth.

"And out of the ground the Gods *formed every beast* of the field, and every fowl of the air." (Abraham 5:20. Italics ours).

Thus, the Gods first created the earth, then planned its development (the spiritual creation). The plan proceeded from the advent of plants to animals and finally to man. However, in the actual process of populating the earth after its creation (the temporal creation), Adam came first, then vegetation was introduced and finally the animal kingdom was brought forth.

How long was Adam on the earth before the fall? That is not known, but it is stated that when Adam received the instruction concerning the tree of knowledge of good and evil, time as we know it was not yet being measured.

"Now I, Abraham, saw that it was after the Lord's time, which was after the time of Kolob; for as yet the Gods had not appointed unto Adam his reckoning." (Abraham 5:13).

So when time is spoken of during this period before Adam was appointed his reckoning, it must be God's time rather than the time scale used by men. Thus, when Adam was advised that if he were to partake of the forbidden fruit *"in the day thou eatest thereof thou shalt surely die."* (Moses 3:17), that day could be interpreted as a day according to the Lord's time, which is equivalent to 1,000 years of earth time. However, the Hebrew word, 'yowm', from which the word 'day' was translated may also be rendered as 'time'. Thus the phrase *in the day thou eatest thereof* (Moses 3:17), could also be interpreted as *in the time thou eatest thereof*. That very phrase is used in Abraham 5:13 – "*for in the time that thou eatest thereof, thou shalt surely die.*" Considering that interpretation of the original 'yowm' there would be no way in which any time scale could be attached to the period when Adam would have transgressed the law given to him in the Garden of Eden prior to his expulsion from the Garden.

Prior to their partaking of the forbidden fruit Adam and Eve lived on the earth as a terrestrial, rather than a telestial, sphere. Some indication of the conditions prevalent on the terrestrial sphere are given by Elder Bruce R. McConkie:

"In its primeval, edenic state all of the earth's surface was in one place (Moses 2:9); thorns, thistles, briars and noxious weeds had not yet begun to grow on it; rather, all plant and animal life was desirable, congenial, and designed to provide for man (earth's crowning inhabitant) a fruitful, peaceful garden in which to dwell. It was not a condition attained by progressive, creative evolvement from less propitious situations; it was creation in its glory, beauty, and perfection; hence, the Lord God pronounced it 'very good'. The fall to present conditions was to come later." (Bruce R. McConkie, Mormon Doctrine, 2d ed., p.251).

We learn from Brigham Young when Adam fell, the earth fell also, and was transformed into a telestial sphere.

> "When the earth was framed and brought into existence and man was placed upon it, it was near the throne of our Father in heaven. And when man fell...the earth fell into space and took up its abode in this planetary system, and the sun became our light. When the Lord said-- "Let there be light," there was light, for the earth was brought near the sun that it might reflect upon it so as to give us light by day and the moon to give us light by night. This is the glory the earth came from, and when it is glorified it will return again into the presence of the Father, and it will dwell there, and these intelligent beings that I am looking at, if they live worthy of it, will dwell upon this earth." (Journal of Discourses 17:143).

So apparently when Adam partook of the forbidden fruit on that terrestrial sphere *the earth fell into space and took up its abode in this planetary system* and became a telestial world. That process could have occurred, according to the meaning intented by the Hebrew 'yown', any time prior to Adam's expulsion from the Garden of Eden.

The Prophet Brigham Young specifically states that Adam was born on another earth and was the chief operator in the development of this world. He records:

"Though we have it in history that our Father Adam was made of the dust of this earth and that he knew nothing about his God previous to being made here, yet it is not so; and when we learn the truth we shall see and understand that he helped to make this world and was the chief manager in that operation.

"He was the person who brought the animals and the seeds from other planets to this world and brought a wife with him and stayed here. You may read and believe what you please as to what is found written in the Bible. Adam was made from the dust of an earth, but not from the dust of this earth. He was made as you and I are made, and no person was ever made upon any other principle."

The planning of this earth and its actual creation were carried out in chiastic order. In the planning stage, plants and animals were created first, followed by the advent of Adam and Eve. In the actual work of creation, Adam came first and was instrumental in the succeeding advent of plants and animals.

This astounding revelation, that the earth is a newcomer to the solar system and that prior to its appearance, time as we know it did not exist on the earth, seals the demise of the time scale on which Darwin's survival of the fittest is based.

So from the word of God we understand that there were two creations of which the scriptures speak -- first a spiritual creation and then a temporal creation. The flora and fauna that adorn this earth were brought from other planets by Adam (and, I would assume, co-workers) in the general state that we now find them. They did not evolve from one species to another by Darwinian process, although they have certainly been changed and modified, within species, in response to changes in environmental conditions.

We are then left with this option -- to believe and accept man-made theories that have projected into the distant past the suppositions of scientists relating to plant and animal life from circumstances theorized from present conditions, or to believe and accept the specific statements of the servants of God who speak authoritatively in His name concerning the creation and the history of the earth and its inhabitants.

The Creation Timeline

The concept of time as a one-dimensional uniform flow of the future into the past, is a concept that is restricted to the telestial earth. Time as we know it is not an intrinsic element of reality; it is a temporary restriction of reality. Time as we measure it was not part of reality until Adam had been given his reckoning.

The temporal creation followed the spiritual creation. It was during the temporal creation that Adam was given the command not to partake of the forbidden fruit. The earth at the time was a terrestrial sphere and a satellite of the star Kolob. President John Taylor made the following statement in this regard, as reported in the Doctrinal Commentary on the Pearl of Great Price, p.143, by Hyrum L. Andrus,

"The fact that the earth was in the presence of God, near unto Kolob, at the time of the creation finds scriptural support in the Book of Abraham. In giving the injunction to Adam not to partake of the forbidden fruit in the garden, the Gods cautioned, '*In the time that thou eatest thereof, thou shalt surely die.*' Abraham then observed: "Now I, Abraham, saw that it was after the Lord's time, which was after the time of Kolob; for as yet the Gods had not appointed unto Adam his reckoning." If the Gods had not then appointed to this earth its present reckoning of time, it follows that until after the fall occurred the earth was on a different time scale -- the time scale of Kolob. In other words, until after the fall the earth was not rotating around the sun in our solar system. As before indicated, it was not then placed under its present temporal law. Instead, it was in the presence of God; and while there, it was identified with Kolob in such a way as to be governed by that mighty sphere's system of time."

The time frame of Kolob is explained in the Book of Abraham, Facsimile No. 2 that states that "one day in Kolob is equal to 1,000 years according to the measurement of this earth". Assuming that the terrestrial earth rotated around

Kolob, we know it was the rotation rate of Kolob, not the revolution rate or rotation rate of the earth in that environment, that would be used to measure time before the fall of Adam. We learn in Abraham 3:4 that the Lord's time was measured according to the revolution rate of Kolob.

> "And the Lord said unto me, by the Urim and Thummim, that Kolob was after the manner of the Lord, according to its times and seasons in the revolutions thereof; that one revolution was a day unto the Lord, after his manner of reckoning, it being one thousand years according to the time appointed unto that whereon thou standest. This is the reckoning of the Lord's time, according to the reckoning of Kolob."

So when the Lord told Adam that in the day (or in the time) that he would partake of the forbidden fruit he would surely die, *Adam had not yet received his reckoning.* Thus the "day" in the Lord's statement could be replaced with "in the same period of time" or "in the same epoch" in which he would partake of the forbidden fruit.

It may not be proper to project even the concept of time as we know it to any period prior to the fall of Adam. The one-dimensional time scale, with the future melting into the past at the interface that we call the present is restricted to the seven thousand years of the earth's *temporal* existence. Following the seven thousand year restrictive period, and one would assume also prior to the fall of Adam, time as we know it does not exist. We learn from Doctrine and Covenants Section 88 that at the time of the resurrection six angels shall stand forth "and reveal the secret acts of men, and the mighty works of God" during each of the six thousand years of the earth's temporal existence. Then "the seventh angel shall sound his trump; and he shall stand forth upon the land and upon the sea, and swear in the name of him who sitteth upon the throne, that there shall be time no longer". (Doctrine & Covenants 88:110).

Apart from the gross inaccuracy of the Potassium-Argon decay rate measurement technique, if indeed time is measured for only a period of seven thousand years, this statement puts a definitive end to all projections of time prior to the fall of Adam. We simply *do not know* when pre-historic events occurred, and there will be no way of finding out by any method which is couched in the man-made concepts of mortality in a telestial world! All scientific method is restricted to concepts that may be perceived and measured only in a telestial environment. The earth's duration as a teletial sphere is for only a period of seven thousand years. Thus it is folly to attempt to project telestial concepts related to the earth into the period when it was a terrestrial sphere, and when it was not even an inhabitant of the solar system. The ultimate vindication of catastrophism over uniformitarianism is the great catastrophic event by which the earth itself was transformed from a terrestrial to a telestial world in relatively recent past.

The Revealed History of the Earth

Modern day prophets have given some clear insight as to how the earth was created, where it came from and in what condition it existed prior to its present state. This knowledge is at variance with the currently held beliefs of the scientific community. The key thing to consider when making comparisons between religious information and scientific theories is that science is capable of dealing only with telestial conditions -- the conditions that exist on this earth during the mortal phase of its existence. It is only natural that, with only the information available from observation and deduction, the scientific community can only extrapolate the observed conditions into the distant past and distant future. However, the nature of the earth has changed from one type of sphere into another, and will yet change again from its present state into a future state vastly different from its present condition. Beyond the points of these changes all scientific information and theory cease to have any value since these fundamental and marked changes in the nature of the earth are beyond the reach of both theory and experiment of the scientific method.

1909 marked the centennial of Charles Darwin's birth and the semi-centennial of the publication of his *Origin of Species*. Seven months after the Darwin Centennial, and perhaps in response to questions raised during the Darwin celebration, the First Presidency of the Church of Jesus Christ of Latter-day Saints, consisting of President Joseph F. Smith and counselors John R. Winder and Anthon H. Lund, prepared a position statement on the origin of the physical man. This position statement appeared in the November 1909 issue of the official *Improvement Era:*

> "...It is held by some that Adam was not the first man upon this earth, and that the original human being was a development from the lower orders of the animal creation. These, however, are the theories of men. The word of the Lord declares that Adam was 'the first man of all men' (Moses 1:34), and we are therefore in duty bound to regard him as the primal parent of our race. It was shown to the brother of Jared that all men were created in the *beginning* after the image of God; and whether we take this to mean the spirit or the body, or both, it commits us to the same conclusion: Man began life as a human being, in the likeness of our Heavenly Father.
>
> "True it is that the body of man enters upon its career as a tiny germ or embryo, which becomes an infant, quickened at a certain stage by the spirit whose tabernacle it is, and the child, after being born, develops into a man. There is nothing in this, however, to indicate that the original man, the first of our race, began life as anything less than a man, or less than the human germ or embryo that becomes a man.

"Man, by searching, cannot find out God. Never, unaided, will he discover the truth about the beginning of human life. The Lord must reveal Himself, or remain unrevealed; and the same is true of the facts relating to the origin of Adam's race— God alone can reveal them. Some of these facts, however, are already known, and what has been made known it is our duty to receive and retain...." (*Improvement Era*, Vol. 8. Nov. 1909).

CONCLUSIONS

Darwin himself expostulated on the limitations of his theory. He said:

> "If it could be demonstrated that any complex organ existed which could not possibly have been formed by numerous, successive, slight modifications, my theory would absolutely break down." – Charles Darwin *The Origin of Species* (1859).

Commenting on this statement by Darwin, Dr. Michael Behe, Professor of Biochemistry, Lehigh University, wrote,

> "To Darwin, the cell was a 'black box' – its inner workings were utterly mysterious to him. Now, the black box has been opened up and we know how it works. Applying Darwin's test to the ultra-complex world of molecular machinery and cellular systems that have been discovered over the past 40 years, we can say that Darwin's theory has 'absolutely broken down'." (Michael Behe, biochemist, *Darwin's Black Box* 1996).

Sir Harold Jeffreys, one of the world's leading geophysicists, after carefully examining the evidence for each of the various theories of how our solar system evolved into existence, summarized the situation in this way:

> "To sum up, I think that all suggested accounts of the [evolutionary] origin of the Solar System are subject to serious objections. The conclusion in the present state of the subject would be that the system cannot exist." (Harold Jefferys, *The Earth: Its Origin, History, and Physical Constitution*, 1970, p. 359).

So it is established and recognized by eminent scientists today that the theory of evolution as developed by Darwin has no basis in truth. The fact that the universities and academia in general hold to the defunct idea of evolution is nothing more than a demonstrated aspect of human nature.

Consistent with this statement, John Kenneth Galbraith, (1908-2006) considered by many as the "Last American Institutionalist", is quoted as saying, "We associate truth with convenience, with what most closely accords with self-interest and personal well-being or promises best to avoid awkward effort or unwelcome dislocation of life. Economic and social behavior are complex, and to comprehend their character is mentally tiring. Therefore we adhere, as though to a raft, to those ideas which represent our understanding." (J.K. Galbraith, *The Affluent Society*, Chapter 2, 1958).

It is human nature to protect one's belief system, to protect one's job, to protect one's reputation. This is no less true among university professors than it is among car salesmen. The university professor, however, is in the more vulnerable position. He is protecting a particular view of the universe, while the car salesman is only protecting a given make of automobile. If the car salesman is wrong, it may only impact in a rather minor way the income potential of a given used car lot, whereas if the professor is shown to be wrong, a world-view of nature may be in jeopardy. *The wise student will take as tongue-in-cheek the entire structure of the current view of reality as taught in the institutions of higher learning.*

Concerning the indeterminate past there is very much that we do not know and that we can never find out using the limited procedures of the scientific method. All knowledge concerning conditions both before and after the temporal, telestial phase of the earth's existence can come from only one source, and that source is God Himself as He communicates that knowledge to those who are susceptible by their righteousness and obedience to correct principles to hear the voice of the Lord. The Lord has given his word on that subject in Doctrine and Covenants, Section 76:

> "For thus saith the Lord -- I, the Lord, am merciful and gracious unto those who fear me, and delight to honor those who serve me in righteousness and in truth unto the end. Great shall be their reward and

eternal shall be their glory. And to them will I reveal all mysteries, yea, all the hidden mysteries of my kingdom from days of old, and for ages to come, will I make known unto them the good pleasure of my will concerning all things pertaining to my kingdom. Yea, even the wonders of eternity shall they know, and things to come will I show them, even the things of many generations. And their wisdom shall be great, and their understanding reach to heaven; and before them the wisdom of the wise shall perish, and the understanding of the prudent shall come to naught. For by my Spirit will I enlighten them, and by my power will I make known unto them the secrets of my will—yea, even those things which eye has not seen, nor ear heard, nor yet entered into the heart of man." (Doctrine& Covenants 76:5-10).

The ultimate purpose of this small treatise is to help those who may be awed by scientific pronouncements and who have been taught to believe that the universities are the bastions of knowledge *to learn to judge the professors by the prophets, rather than judging the prophets by the professors.*

ADDENDUM

PHI

The Piltdown man was touted as the missing link in the theory of evolution between the near hominid and man. Discovered in the Piltdown quarry in Sussex, England, by an amateur fossil hunter, Charles Dawson (not Darwin) in 1912, it was identified in 1953 to be a hoax. So, one of the established stepping stones in the evolutionary process between ape and man dissolved. But the lack of transitional forms between ape and man is not the only missing link. Transitional forms between many other species are also missing. In fact, there are enough missing links in the evolutionary sequence between the most primitive identified forms and man to make a substantial chain.

Evolutionists also have a major problem with their starting point -- the generation of living forms. A few amino acids, the building blocks of protein, have actually been generated in the laboratory by the passage of an electric discharge through atmospheric gases containing ammonia, water vapor and methane. Amino acids exist in both right and left versions, and are produced in a 50-50 ratio by an electric discharge. However, living mechanisms use only the 'left' version of the amino acids in their protein, and the simplest protein requires about 250 amino acids in its make-up. Thus, for a living protein to be produced, an extremely improbable series of events must occur -- the spontaneous linking up of some 250 'left' amino acids from a 50-50 mix of both right and left amino acids. Next, this complex protein must somehow be endowed with what we call life -- it must somehow generate the capability to spontaneously reproduce.

The generation of a simple protein is just the beginning of the evolutionary process, and the probability of such a process spontaneously occurring is remote to the vanishing point. On the other side of the argument, if an extremely improbable condition were found to occur prolifically in essentially all life forms, such a circumstance would be a powerful evidence of an intelligent intervention in the creation process. Such a condition indeed exists, and it is associated with a little known mathematical proportion identified as phi. Phi is an irrational number approximately equal to 1.618034, in the same sense that Pi, the ratio of the diameter to the circumference of a circle, is an irrational number approximately equal to 3.141592.

The number Phi was first identified by the ancient Greeks as the height-to-length ratio of the rectangle that was judged to be the most esthetically pleasing, and it was used extensively in their architecture. Such a rectangle was called the Golden Rectangle. The Parthenon, perhaps the most famous building in the world, was constructed with its various dimensions according to this ratio. Michael Angelo, copying the Greek esthetics, used the Golden Rectangle proportion as a dimensional framework for his statue, David.

The Golden Rectangle is of such proportions that if the height were laid off along the base and a vertical erected at that point to make a square; the remaining rectangular portion would have the same length to height ratio as the original rectangle. Thus, that smaller rectangle could be treated in the same way to form a square and a remaining rectangular portion with the same length to height ratio as the larger rectangle -- etc. ad infinitum. That length to height ratio is found to be exactly (1 + 2½)/2, and is called the Golden Ratio.

An independent manner of developing phi was arrived at by Leonard Fibonacci of Pisa (1170-1250). He generated a mathematical series, the first two digits being 0 and 1, then each succeeding digit was the sum of the preceding two, e.g. 0 1 1 2 3 5 8 13 21 34 55 89 144...

The Golden Ratio is derived by dividing any number by its predecessor, e.g., 144/89 = 1.6179775... The first few ratios are not precise, but as this process progresses the succeeding ratios approach progressively nearer to phi until in the limit the exact number is reached. This series has been studied extensively. In May of 2000, for instance, the series was taken out to 1.5 billion decimal places by computer. The process required nearly three hours of computer time.

The degree to which this number, phi, is found in nature is astounding. The following citations are by no means exhaustive, but merely exemplary of the diverse, unrelated fields in which phi is a prominent characteristic.

BOTANY

"Sneezewort", or Achillea Ptarmica
As this plant grows and branches out, the number of successive branches increases according to the Golden Ratio.

Petals on Flowers
On many plants, the number of petals is a Fibonacci number: lilies and iris have 3 petals: buttercups have 5 petals; some delphiniums have 8; corn marigolds have 13 petals; asters, black-eyed susan, and chicory have 21 petals; plantain and pyrethrum have 34; michaelmas daisies and the asteracea family have either 55 or 89 petals.

Seeds on Flower Heads
Fibonacci numbers can also be seen in the arrangement of seeds on flower heads. The seeds are normally arranged in spiral patterns, the number of spirals are frequently the Fibonacci numbers.

Pine Cones
The segments of pine cones have a spiral configuration. Again, the number of spirals are numbers in the Fibonacci series.

Leaves on Plant Stems
Leaves on plant stems are generally offset from one another so as to allow maximum exposure to sunlight. In some plants if one draws a straight line down the stem of the plant, and

counts the number of successive leaves as the stem is rotated on full turn, the number of leaves in each rotation corresponds to the Fibonacci numbers.

BIOLOGY

Snail Shells
If one draws a line from the center of a snail shell to the edge, and the line crosses at least two segments of the shell, the relative widths of succeeding segments will be approximately 1.62.

The Dolphin
The eye, fins and tail of the dolphin all fall at golden sections of the length of the dolphin's body.

Angel Fish
The length of the body of the angel fish is a golden ratio of length from the nose to the tip of the tail. The distance from the nose to the end of the body is a golden ratio of the distance from the nose to the center of the fins. The distance from the nose to the end of the gills is a golden ratio of the distance from the nose to the center of the fins, and the distance from the nose to the center of the eye is a golden ratio of the distance from the nose to the end of the gills.

Penguins
The eyes, beak, wing and key body markings of the penguin all fall at golden sections of its height.

Ants

The body sections and leg sections of the ant are also golden sections of the length of the body.

LIFE FORMS

(As can be found in the website www.goldennumber.net).

The Human Body

The human body is proportioned according to the Golden Ratio. Starting from the head of an erect person, the distance to the bottom of the hands is the golden ratio of the height of the person. The distance from the head to the navel and elbows is the golden ratio of the distance from the head to the bottom of the hands. The distance from the head to the pectorals and the inside top of the arms, as well as at width of the shoulders and the length of the forearm and shin bone is the golden ratio of the distance from the head to the navel and elbows. The distance from the head to the base of the skull, as well as the width of abdomen is a golden ratio of the distance from the head to the pectorals and inside top of the arms.

Another interesting relationship of the golden section to the design of the human body is its relationship to the number 5, one of the prime numbers in the Fibonacci series. There are five appendages to the torso-the arms, legs and head. There are five appendages on each of these-five fingers, five toes and five apertures in the head. (Interestingly enough, the golden section is based on the number 5, i.e., 5 raised to the 0.5 power times 0.5 plus 0.5 is exactly equal to phi).

DNA
The DNA molecule, the program for all life, is based on the golden section. It measures 34 angstroms long by 21 angstroms wide for each full cycle of its double helix spiral. 34 and 21, of course, are numbers in the Fibonacci series and their ratio, 1.6190476, closely approximates phi, 1.6180339...

Music
Consider the 'octave'. It is made up of 13 chromatic tones (including the octave). Let the semitone (1) and the whole tone (2 semi-tones) be the building blocks of the diatonic scale. Pentatonic scales are 5 tones. Diatonic scales are 8 tones. The 1st, 3rd, and 5th tones in the diatonic scale are the building blocks of root chords. All these values are Fibonacci numbers.

Astronomy
For golden ratio proportions in the relative distances of the various planets from the sun and the golden ratio relationship between their various orbits, please access www.solargeomety.com .

THE BIBLE

The Ark of the Covenant

The Ark of the Covenant was constructed according to the Golden Ratio. The instructions for the dimensions of the ark are given in Exodus 25:10, as

> "And they shall make an ark of shittim wood: two cubits and a half shall be the length thereof, and a cubit and a half the breadth thereof, and a cubit and a half the height thereof."

The ratio of 2.5 to 1.5 is 1.666, which is the number Phi to the given numbers of decimal points.

Noah's Ark

The ends of Noah's ark were also constructed to this same ratio:

> "Make thee an ark of gopher wood; rooms shalt thou make in the ark, and shalt pitch it within and without with pitch. And this is the fashion which thou shalt make it of: The length of the ark shall be three hundred cubits, the breadth of it fifty cubits, and the height of it thirty cubits (Genesis 6:14-15).

The ratio of 50 to 30 again is 1.666, the same ratio as the length to the sides as the Ark of the Covenant.

666 and the Colors of the Tabernacle

Both the number 666 (Revelations 13:18) and the colors of
the tabernacle (Exodus 26:1) are based on the Phi
relationship. (See www.goldennumber.net/bible.htm).

CONCLUSION

Such an astounding prevalence of the ratio phi, in essentially all aspects of the world around us, from the DNA molecule to the solar system, makes an undeniable argument for the essential presence of an intelligent being in the creation process. We should stand in awe in contemplating creation, from the simplest life forms to the majesty of the heavens.

> "The earth rolls upon her wings, and the sun giveth his light by day, and the moon giveth her light by night, and the stars also give their light, as they roll upon their wings in their glory, in the midst of the power of God...and any man who hath seen any or the least of these hath seen God moving in his majesty and power." (D & C 88:45-47).

In spite of the prevailing opinions of dogmatic science, that touts itself as being purely objective, the theory of evolution will follow all the previous erroneous notions of scientific truth into the dustbins of history. There are yet a number of questions relating to paleontology and to the history of the earth that are not solved by the theory of catastrophism and are not revealed in the scriptures. We simply have no timeline by which to date any prehistoric life form. We do understand, however, that the dating techniques of modern paleontology simply have no credence.

Fossil fuels, principally coal, petroleum and natural gas, may not all be related to fossils. There is no question that coal was formed from hydrocarbon material, as in many cases the shapes of the original organic material are retained in the coal structure. But petroleum is another matter. Velikovsky, for instance, postulates that much of our petroleum deposits resulted from the transfer of material from Venus to the Earth during one or more of the near passes of the two planets. Velikovsky hypothesized, for instance, and it was confirmed that Venus does have a heavy hydrocarbon atmosphere.

We may be confident that life forms found in deeper strata predate life forms found above them in the same strata, but assigning the same age to the same types of strata in different regions of the earth is a tenuous theory at best. And attempting to date the fall of Adam with the advent and demise of any species simply has no foundation. However, we do know from the scriptures, as mentioned earlier in the text, that Adam's appearance on the earth predated the appearance of all other life forms. But not to be discouraged -- the answers to all of our concerns will be made available if we look in the right places and follow correct procedures. God is the author of all truth, and He has promised that in his own time He will reveal to us all truth.

> "For thus saith the Lord—I, the Lord, am merciful and gracious unto those who fear me, and delight to honor those who serve me in righteousness and truth unto the end. Great shall be their reward and

eternal shall be their glory. And to them will I reveal all mysteries, yea, all the hidden mysteries of my kingdom from days of old, and for ages to come, will I make known unto them the good pleasure of my will concerning all things pertaining to my kingdom. Yea, even the wonders of eternity shall they know, and things to come will I show them, even the things of many generations. And their wisdom shall be great, and their understanding reach to heaven; and before them the wisdom of the wise shall perish, and the understanding of the prudent shall come to naught. For by my Spirit will I enlighten them, and by my power will I make known unto them the secrets of my will—yea, even those things which eye has not seen, nor ear heard, nor yet entered into the heart of man." (D & C 76:5-10).

THE END

H. Clay Gorton - (1923-2008)

Henry Clay Gorton was born on March 7, 1923. Clay devoted his life in service to his Savior. He fulfilled six missions, including service as a mission president in Cordoba, Argentina and as the president of the Missionary Training Center in Santiago, Chile. He also served as a stake president, as a bishop, as a welfare services region agent for the Los Angeles, California region, as a counselor to four stake presidents, and as a member of three high councils. His final calling was as a sealer in the Bountiful, Utah Temple.

Following his first mission, he obtained a Master of Arts degree in physics and mathematics at Brigham Young University. After graduation, he conducted applied research in semiconductor physics at Battelle Memorial Institute in Columbus, Ohio for sixteen years. After his service as mission president, he joined the staff of the Electronics and Defense Sector of TRW in Redondo Beach, California. Following his retirement from TRW in 1986, he served as Chief Scientist at TRW Components International.

He had been active in the Institute of Electrical and Electronic Engineers, The Institute of Environmental Sciences and The Electrochemical Society. He served as Divisional Editor for the Journal of the Electrochemical Society. He is the author of eighteen technical publications, and has been a guest speaker for the American Society for Quality Control.

He has also been a guest lecturer at the Moore School of Engineering in Philadelphia, Pennsylvania.

After moving to Bountiful, Utah in 1992, he obtained his pilot's license and enjoyed aerobatic flying. He served four years as the editor for the Salt Lake Chapter of the Experimental Aircraft Association (EAA) newsletter and three years as the editor of the Starduster Magazine.

Clay authored eight books (including this one) and has written a number of papers on religion, philosophy, health and aviation. His last years were spent serving as "Gramps" on www.askgramps.org, where he adopted thousands of people worldwide as his "grandchildren", assisting them through his wit, wisdom and sound advice.

His exemplary life has influenced many people around the world.

Other Titles by H. Clay Gorton

LANGUAGE OF THE LORD
Horizon Publishers, August 1993, 351 pages.
New discoveries of Chiasma in the Doctrine &
Covenants.

THE LEGACY OF THE BRASS PLATES OF LABAN
Horizon Publishers, November 1994, 298 pages.
A comparison of Biblical and Book of Mormon Isaiah
texts.

A NEW WITNESS FOR CHRIST
Horizon Publishers, March 1997, 489 pages.
Chiastic structures in the Book of Mormon.

ASK GRAMPS, VOL 1
Maasai, Inc. Provo, Utah, 2001, 134 pages.
Addressing 101 Everyday Concerns, Curiosities, and
Uncertainties of Latter-Day Saints.

ASK GRAMPS, VOL 2

Maasai, Inc. Provo, Utah 2002, 128 pages.

Another 101 Everyday Concerns, Curiosities, and Uncertainties of Latter-day Saints.

ASK GRAMPS FOR TEENS

Maasai, Inc. Provo, Utah, 2002, 125 pages.

What is Really on the Minds of Young People Today.

MAN'S ETERNITY: HIS JOURNEY, HIS DESTINY

Maasai, Inc. Provo, Utah, 2002, 154 pages.

A Treatment of Frequently Asked Questions about the Plan of Life and Salvation and the Basic Principles of the Gospel of Jesus Christ.